French transporter. The towers are 204 feet high and the bridge, 530 feet long, is 150 feet above the water. The car is 16 feet above the river at high tide, and the trolley was originally driven by cables worked by a steam engine in the east tower.

In 1939 the bridge was converted to electric traction and the speed increased, being now approximately 300 feet per minute. Fares are charged, and there is quite a large range of categories: passengers, goods, animals, carts, cars, lorries, cycles, funeral parties and night journeys.

France

Rouen: The nearest French transporter bridge to Britain, and the first one to be opened, was that across the Seine at Rouen. It began service on September 16, 1899, and had towers 221 feet in height, the span being 164 feet above the water and 474 feet in length. It was a suspension bridge. The car, 42 feet long by 36 feet wide, hung 140 feet below the span. The crossing took 55 seconds and the car was electrically driven, the winding equipment being mounted on the car, which thus winched itself to and fro across the river. This arrangement, which added about seven tons to the car's weight, was not repeated. Members of the public were allowed to ascend the towers for sightseeing purposes, the charge in pre-1914 days being 50 centimes, about 5d. There were first and second class shelters, at fares of 10 and 5 centimes respectively. Tramway routes existed on both banks of the river and it was proposed at one time to run trams across the bridge, but this suggestion was not carried out.

Many stories are told of the Rouen bridge: the diver who won a bet by diving from the top of the bridge into the Seine, the aviator who flew through the opening in 1915, the testing of early parachutes, the submergence of the car in the 1910 floods, and the firework displays. The bridge was blown up by French Army engineers in June 1940 after a life of just over forty years.

On the left: The world's first transporter bridge, designed by Arnodin and built in 1893 at Portugalete, near Bilbao (Spain); it is still in use. *(Libreria E. Verdes*

On the right: A close-up of the 1906 transporter at Newport, Monmouthshire, looking west. The trolley is just visible in the lower right-hand corner. This bridge is still in use, but its future is uncertain. *(Commercial postcard*

Martrou: Next in chronological order was the bridge at Martrou, crossing the Charente not far from Rochefort. It was similar in general design and appearance to those at Portugalete and Rouen. It stood in the middle of open country and use of it was free, the cost being borne by the *Département*. The bridge was opened in 1900, being operated at first by a stationary steam engine but later by electricity. It had a span of nearly 460 feet and was 164 feet above the water. It survived till February 1967 and was the last of the French *transbordeurs;* a new lifting bridge built quite close to it now performs its task. The two bridges, old and new, were portrayed recently on a French postage-stamp of value 25 centimes. The new bridge is in the foreground, raised to allow the passage of a ship, and the transporter is visible beyond. There have been suggestions that the disused bridge should be retained intact as an engineering monument.

Nantes: Arnodin's next transporter was as Nantes, crossing the Loire. This differed from the previous three, being a combination of suspension and cantilever construction. The height of each of the towers was 250 feet; the bridge was approximately 500 feet long and 164 feet above the water. It was driven by electricity, stationary 10-hp motors operating a cable attached to the trolley. The crossing took 60 seconds. The bridge was opened on October 22, 1903, and was in operation until 1957; this and the bridge at Martrou were the only French transporters which survived World War II. On completion, the bridge was tested by loading the car with 50 tons of granite sets.

Marseille: Of similar design to the *transbordeur* at Nantes, but slightly larger, was the one crossing the Vieux Port at Marseille. Its span was 537 feet and its height 164 feet; the towers reached 282 feet. It was one of the chief sights of the city, and survived until 1944. The Germans, wishing to block the entrance to the Old Harbour, mined the towers, but only one of the charges exploded, cutting the bridge in two. Later the remains were demolished. The method of operation was similar to that of the bridge at Nantes.

Brest: The last French transporter bridge to be opened was that at Brest, in 1909. In fact, however, this bridge was secondhand, for it was originally erected at Bizerta naval base in Tunisia in 1898, and opened in the summer of that year. It thus really belongs to the overseas section but, as Tunisia was at the time a French Protectorate, it may be reasonable to describe it here. At Bizerta it crossed an artificial channel, 300 feet wide, connecting a lake anchorage with the sea. The towers were 355 feet apart and the span was 152 feet above the water. The bridge was of suspension type, like the other early ones. The trolley was worked by a cable operated by a stationary steam engine. The car was 30 feet by 23 feet, a comparatively small size. In November 1898, a few months after its inauguration, it managed to stand without damage a severe cyclone which visited the area.

Its life at Bizerta was short, however for after less than a decade the width of the channel was increased and the bridge dismantled. It was given a new lease of existence at Brest, where it was opened across the naval harbour in 1909 electricity replaced steam as the motive power. It is said that a main reason for the removal of the bridge from Bizerta was that, owing to its height, it would provide an excellent target for gunfire from an enemy fleet! At Brest the naval harbour is a cleft-like gorge and this objection did not arise. The bridge was damaged beyond repair in the 1944 fighting and was dismantled in 1947.

There were other proposals for transporter bridges in France. One would have been at Tancarville, over the Seine below Rouen. Another, which would presumably have been the largest in the world, proceeded beyond the projection stage; it would have crossed the Garonne about two miles below Bordeaux. The writer remembers seeing the huge towers in the distance, in 1929. Construction was held up by lack of funds; subsequently it was decided that a transporter would not meet increased traffic requirements and the towers were eventually demolished during World War II. A high-level road bridge has recently been erected at an adjacent point.

Germany

The only other European transporters were built in what is now West Germany, and two of them still survive, at Osten and Rendsburg.

Kiel: Early in the present century an extension was made to the naval harbour at Kiel, beyond a deep inlet; to maintain communication with this, a transporter bridge of suspension type was built across the mouth of the inlet. It was opened in 1909 and had the shortest life of any of the world's trans-

Above: The Newport transporter car, still in use today and scarcely altered. The safety-net appears in several early views, but whom did it protect, and from what?
(Philco Series

Centre: The cantilever-type transporter at Nantes, built in 1903 and dismantled in 1957.
(F. Chapeau

Below: The transporter car at Brest, which was in use from 1909 to 1944. The bridge was originally erected at Bizerta in 1898 but later removed.
(Neurdein et Cie

porters. After World War I the installations beyond the inlet were no longer in use; the bridge was an obstacle to shipping entering the inlet and was also in need of repair, so it was closed in 1923. The inlet has been filled in, and so the whole area now consists of level ground. The span of the bridge was approximately 420 feet and its height 160 feet.

Osten: In the same year, 1909, a transporter bridge was opened at Osten, a small village about 30 miles north-west of Hamburg. It is known as the *Schwebefähre Osten-Basbeck* and crosses the Oste, a small river which flows into the Elbe estuary not far from its mouth. *Schwebefähre,* the German word for a transporter bridge, means literally "hovering ferry", and the first part of the word will be familiar to all readers of this journal through the famous *Schwebebahn* at Wuppertal.

The Osten transporter is on a fairly small scale, and is a rather ugly structure of the truss type. It has a span of about 260 feet and its height above the Oste is 114 feet. A bus service running once daily in each direction between Bremen and Itzehoe crosses the bridge. The car is suspended by rigid girders instead of cables as in the other bridges so far described.

Rendsburg: At Rendsburg in Schleswig-Holstein a high-level railway bridge, with very lengthy approach viaducts, crosses the Kiel Canal. This bridge is of cantilever construction and, on the underside of this bridge, there was opened in 1913 a transporter to serve the needs of road traffic. The span is approximately 140 feet above the canal and its length is 480 feet. Recently a tunnel was opened for motor traffic, but this is about a mile away; the transporter remains in use for pedestrians, cyclists, the occasional horse-drawn vehicle and such motorists as may not wish to use the tunnel. The car, like that of the Osten bridge, is suspended by girders, and radar is fitted.

Transporter bridges overseas

The three transporter bridges which have existed outside Europe are that at Bizerta, already described, one in the U.S.A. at Duluth, Minnesota, and a third at Rio de Janeiro in Brazil.

Duluth: The transporter bridge in this city was remarkable in being relatively short: a 394 feet span with a height of 135 feet above the water. It crossed the narrow canal separating the harbour from Lake Superior. The steel truss type of construction was used, and the car was suspended by a fixed framework as in the case of Osten. The bridge was opened in the Spring of 1905. In its construction special allowance had to be made for the climatic extremes which prevail in the Middle West of America. The trolley ran on 32 wheels and the car measured 34 by 50 feet, having two enclosed cabins for passengers and a space between them large enough for six motor cars. It was worked by cables operated by two 40-hp motors, and each round trip took 13 minutes.

With the passage of time and the increase in road traffic, both vehicles and shipping were being delayed by the bridge, and in 1927 the city made plans for the conversion of the structure into a lift bridge. The work began in 1929 and the bridge in its new form was opened in May 1930. It is still in use, and still retains the odd appearance that it had in comparison with other transporters, the highest point being the middle of the span.

Rio de Janeiro: The Brazilian transporter, the *Ponte-Pênsil Alexandrino de Alencar,* was opened in February 1915, thus being the last public transporter bridge to be completed in the world. Its construction, however, entrusted to a German firm, had lasted for more than four years, having begun in December 1910. It joined the mainland to an island, Ilha das Cobras, on which there was a naval installation.

The bridge was of the suspension type and is described as having been built according to the *sistema Runcorn.* It had a span of approximately 560 feet and was about 65 feet above the water. Motive power was by 2 x 36-hp motors mounted on the trolley itself. The car, suspended by rigid girders, was comparatively small. The bridge had a comparatively short life, being closed in 1935 and replaced by a fixed bridge, the *Ponte Arnaldo Luz.*

Transporter bridges in Britain

Our own country has had a total of four transporter bridges, three of which still survive. Each of them differed considerably from the others. They will be dealt with in the order of their construction.

Widnes: This bridge over the River Mersey and the Manchester Ship Canal is sometimes known by the name of Runcorn, the town on the Cheshire side of

Above: Most transporter bridges were of the suspension type, but those at Marseille, Nantes and Middlesbrough were built on the cantilever pr.nciple. The Marseille bridge, shown here, had a passenger lift in the north tower, giving access to the trolley-way and a restaurant; the bridge was demolished in 1944.
(Phototypie Lacour

Centre: The 1911 transporter that links Middlesbrough with Port Clarence, across the River Tees. *Transporter* is a familiar destination on the local buses, and was formerly on the trams.
(Rotary Photo Co

Below: As originally built, the Rouen transporter was unique in having its electrically-powered winding gear mounted on the car, which thus winched itself to and fro across the Seine. The equipment was later moved ashore to lighten the car, and plans were drawn up to run trams across the bridge, but this did not materialise. The Rouen bridge was blown up in June 1940.
(Imprimeries Réunies

Above: The transporter bridge across the Mersey from Runcorn to Widnes, opened in 1905 and dismantled in 1961. It was unusual for its relatively low height and for its length; the crossing time was three minutes.
(W. R. Hall, Widnes

Second: A little-known transporter was that in the naval base at Kiel (Germany), built in 1909 and dismantled in 1923.
(Verlag H. Edlefsen, Kiel

Third : Of the seven transporters still existing, the least well-known is probably the private bridge of Joseph Crosfield & Sons across the Mersey at Warrington, which can take both road and rail vehicles. Although no longer in regular use, the bridge is maintained in working order and can be demonstrated to visitors.
(A. R. Phillips

Below: At Rendsburg, in North Germany, the high level railwa bridge across th Kiel Canal in cludes a transpor ter car suspende from the rail deck brought into us when the cana was widened i 1913. This photo graph by Ger Wolff shows a die sel train crossin above the car.

the river, but it belonged to and was operated by Widnes Corporation. It was probably the best known of all transporters, and it was certainly the longest, with a span of 1 000 feet. The fact that it was considerably less high than most others, 82 feet above the water, made it look longer still. A greater height than this was not necessary, for the Mersey is not navigable by large craft at this point, and the Ship Canal was already subject to a height restriction owing to the presence of several other bridges, including the Runcorn railway bridge a short distance below the transporter.

The Widnes transporter, which was opened in May 1905, was of the suspension type. The four towers, each 190 feet high, were of a particularly neat design, and the main cables were each a foot in diameter, their total weight being 243 tons. They were anchored at each end in holes bored to a depth of 30 feet. The span consisted of two girders 18 feet deep and 35 feet apart. The trolley, 77 feet long, was slung from the wheels, 32 in number, instead of resting on them, and it picked up current for its two 35-hp motors from conductors running the length of the bridge. The car was suspended by cables, which, in later days at least, used to rub against one another as it was crossing. It was 12 feet above high water level in the river, and passed 4 feet 6 inches above the embankment separating the river from the Ship Canal. It was 55 feet long and 24 feet wide, with covered accommodation for foot passengers on the seaward side. The crossing took two and a half minutes, and the car had magnetic brakes. In 1913 the drive was changed to stationary motors winding a cable on to a drum.

Fares were charged for pedestrians and vehicles, the rate during the last years being 2d for the former. One could cross the river for 1d by walking across the adjacent railway bridge, but for vehicles other than pedal cycles there was no alternative to the transporter. With increasing traffic the bridge became inadequate for the needs of the district, and in July 1961 a new arch bridge, a smaller version of that over Sydney Harbour, was opened, the transporter being closed with an official ceremony on the following day, and subsequently demolished. The new bridge is toll-free, and tolls on the footway across the railway bridge were abolished at the same time. The footway, however, has since been closed.

The Widnes bridge filled a much more vital need than either of the other two public transporters in Britain, for anyone in a vehicle not wishing to use it had to make a detour of many miles: either to a swing bridge at Stockton Heath, Warrington, six miles upstream, or to the ferries (or from 1934 the Mersey Road Tunnel) thirteen miles downstream between Liverpool and Birkenhead. The other two, at Newport and Middlesbrough, are in each case only about one mile downstream from other road bridges, and therefore handle only traffic of a purely local character.

Newport: The transporter bridge at Newport, Monmouthshire, closely resembles Arnodin's suspension bridges of the original design — Portugalete, Bizerta, Rouen, Martrou. It was opened in 1906, and crosses the Usk at a spot at which a tunnel was proposed in 1888, but this was not built. A ferry operated here until rendered redundant by the opening of the bridge. The span is 645 feet long and 177 feet high; it has 16 suspension cables, each composed of 127 wires, and the breaking weight of a cable is some 254 tons, giving a very ample margin of safety.

The trolley, 104 feet long, runs on 60 wheels. The car is 33 feet long and 40 feet wide, with covered accommodation for foot passengers on either side of the space for vehicles. It is propelled by two motors, each of 35-hp; these drive a drum on which a continuous steel cable is wound. Since 1946 no charge has been made for the use of the bridge, which is operated by Newport Corporation, except in the case of persons wishing to climb the towers and walk across the structure for sightseeing purposes; present "fare" for this exploit is 6d.

Unfortunately the future of this bridge is by no means certain. Newport Corporation regard it as a "white elephant", and would like to close it. It is costing about £15 000 a year to run, but its demolition would demand at least £100 000. There is a possibility that the bridge might be bought by an American, Mr Douglas C. Jones, for re-erection in the USA as a curio. But at the moment no alternative way of crossing the river at this point has been established.

Middlesbrough: The Middlesbrough—or now, more strictly, Teesside—transporter crosses the River Tees to Port Clarence in County Durham. It is of cantilever construction and has thus quite a distinctive appearance. It was opened in

Above: The only transporter in North America was the Aerial Ferry Bridge across the ship canal at Duluth, which carried pedestrians, road vehicles and trams. It was in use from 1905 to 1927, and was then converted to a lifting bridge.
(courtesy Robert O. Brown

Below: Similar in construction to the Duluth bridge, but somewhat lighter, is the *Schwebefähre* across the River Oste at Basbeck-Osten, between Bremen and the Elbe.
(J. H. Price

October 1911, and it seems reasonable to assume that it will be the last such bridge to survive in Britain, and perhaps in the world. In fact, early in 1969, workmen placed pipes across the span to transfer oxygen between two Teesside industrial plants, and the bridge has recently been repainted.

The length of the span is 570 feet and its height above the water 140 feet. It is operated by electric motors winding a cable on to a drum. Official figures concerning the dimensions include the statement: *"Tested with* 80 *tons of pig iron. Guaranteed to carry* 860 *passengers,* 600 *with Tramcar and passengers".* It would appear, therefore, that there must have been at the time of its opening a plan to build an extension to the tramways on the Port Clarence side of the river, but nothing ever came of this. Tolls are charged for the use of the bridge.

Crosfield's, Warrington: Comparatively unknown is the private transporter bridge across the River Mersey at the extensive soap and chemical works of Messrs Joseph Crosfield & Son Ltd, a Unilever subsidiary, at Warrington. These works are on both sides of the river, and for many years the bridge provided access to the south side for both road and rail vehicles, it being to the best of the writer's knowledge the only transporter with rails laid on the floor of the car.

Compared with other bridges described, this one is not a large structure. It is of cantilever type, the span being 187 feet long and 75 feet above the water. It is cable worked by an electric motor. Since 1964 it has not been in regular use, a new road approach having been constructed to the part of the works south of the river, but it is maintained in perfect working order for possible further use in the future. It was opened in 1916, its construction having been delayed by World War 1.